A Da[illegible]
Life of Oakdale

by Kris Carlin
illustrated by Dani Jones

SCHOOL PUBLISHERS

Printed in China

ISBN 10: 0-15-350782-9
ISBN 13: 978-0-15-350782-3

Ordering Options
ISBN 10: 0-15-350601-6 (Grade 4 On-Level Collection)
ISBN 13: 978-0-15-350601-7 (Grade 4 On-Level Collection)
ISBN 10: 0-15-357919-6 (package of 5)
ISBN 13: 978-0-15-357919-6 (package of 5)

2 3 4 5 6 7 8 9 10 985 12 11 10 09 08 07

Harcourt

Characters

Narrator, Leanne, Miss Greene, Gordy
Charlie, Caitlin, Frieda, Mom
Lucy, Dad, Franklin, Brian

Setting: A fourth-grade classroom

Narrator: It's time for social studies in Miss Greene's fourth-grade class.

Miss Greene: Does anyone know what a time capsule is?

Charlie: Is it what makes watches run?

Frieda: Is it a type of medicine that takes a while to work?

Lucy: I think it has something do with time travel. It could be a type of spacecraft that's able to travel into other times and dimensions.

Narrator: Everyone gapes at Lucy with some confusion and consternation.

Lucy: I can dream, can't I?

Miss Greene: Those are all very good guesses. However, a time capsule is something different. It's a collection of things that tell about the past.

Franklin: Like a museum?

Miss Greene: Sort of—museums do indeed keep extensive collections of things from the past. You could also say a library contains the past in its books. The attic of your house holds the past. A time capsule, though, is a particular collection of things that is specifically set aside so that people in the future can learn about everyday life in another time.

Leanne: We're going to create a time capsule to show fourth graders in the future what we did and what we liked?

Miss Greene: Exactly.

Gordy: What should we put in it?

Miss Greene: That's your homework. Find an item to put into our time capsule. Choose something that tells about your life and the world today.

Caitlin: Can we really bring in anything?

Miss Greene: Absolutely—anything except food, that is.

Narrator: The bell rings.

Miss Greene: Have a good weekend! Remember to bring something Monday for the time capsule!

Narrator: Brian trudges into the kitchen of his family's house. He drops his backpack to the floor and slumps into a chair at the kitchen table. His younger sister Stacey sits at the table, serenely drawing, as Mom talks on the phone.

Mom: All right, I'll see you at noon on Monday. Bye!

Narrator: Mom flips her phone shut and sits down across from Brian.

Mom: You're looking unusually downcast for a Friday afternoon. What's going on?

Brian: We're making a time capsule in social studies, and I have to bring something to put into it.

Mom: That sounds like fun. What are you going to bring?

Brian: I don't know. That's why I don't think it sounds like fun.

Mom: Hmm . . . a time capsule has to tell people in the future about life today. What are things you do? What are things you like? How about pizza?

Brian: No food allowed!

Mom: How about a video game?

Brian: I'm not depriving myself of one of my games. Besides, I bet others will have that idea. I want something special, something that will stun and amaze the class.

Mom: Well, it should be something that will be reminiscent of our time. Why don't you keep track of everything you do this weekend? Then on Sunday, look back and decide what best represents you.

Narrator: It's early Saturday. Brian is awakened by his dog, Skipper. He averts his face from the wet kisses and then gets up. He mutters to himself as he stumbles around the room.

Brian: Okay, Skipper. Let's go for a walk. Oh, wait, I'm supposed to keep track of everything I do today. Where's my notebook?

Narrator: Brian looks around his room for his notebook. Instead, he sees his camera lying on his desk. He stares at it pensively for a moment.

Brian: I think I have a better idea. Come on, Skipper!

Narrator: Brian and Skipper run outside. As they stand on the front steps, Brian snaps a picture of the leaping Skipper.

Narrator: Later that morning, Brian finds Mom and Dad in the kitchen.

Brian: Mom, could I recruit you to help with my time capsule project?

Mom: What do I have to do?

Brian: All you have to do is let me come to the grocery store with you.

Mom: I think I can handle that.

Brian: Also, Dad—can I come with you later when you take Stacey to karate?

Dad: I think you're a little big for her class.

Brian: I'm not taking the class. I just need a little everyday Oakdale.

Narrator: Now it's Monday, and it's time for social studies. Miss Greene's class is outside by a big rock with a plaque on it that reads "Oakdale Elementary School." Miss Greene holds a shovel and stands near a hole in the ground. There is a metal box at her feet.

Miss Greene: Principal Sheridan asked the custodial staff to dig a hole where we'll bury our time capsule. I've left instructions with the school board and the town hall for the fourth-grade class that will be here in twenty years. They'll dig up the time capsule and learn all about you. Now, let us commence! Who wants to start?

Gordy: I'm first! I have a collection of boxes from my favorite video games.

Frieda: I brought a copy of today's newspaper.

Charlie: This was the most popular movie of the year.

Leanne: You said no food, but I thought we should have a culinary exhibit, so here's a copy of our cafeteria menu. It has a plastic cover to keep it in good condition.

Caitlin: Here's one of our school T-shirts.

Lucy: I made an imprint of my hand in plaster. I also put some nail clippings and hair in this bag so that kids in the future can examine my DNA.

Class: Yuck!

Narrator: Miss Greene looks slightly queasy but manages a smile as she takes the bag from Lucy.

Miss Greene: I think that's a very interesting idea. Let's put everything in the box.

Narrator: Miss Greene puts the items into the box. Brian then steps forward, holding a three-ring binder.

Miss Greene: What do you have, Brian?

Narrator: Brian holds up the binder. The title on the cover is "An Oakdale Saturday."

Brian: I took pictures all day on Saturday and put them together into a photo album.

Narrator: The class gathers around Brian. He shows them the picture of Skipper in the early morning, and then a picture of his neighbor picking up a newspaper at the end of the driveway. Times are printed under each photo.

Caitlin: Hey! I'm in this picture! There I am with my mom outside the grocery store!

Narrator: They see pictures of people picking out food or waiting in line to check out.

Franklin: That's my karate school!

Narrator: They look at pictures of a karate class as well as pictures of students leaving the class with their parents.

Brian: Here's the park on a Saturday afternoon.

Narrator: He shows them photos of people jogging, dogs running and playing, and soccer games in progress.

Brian: Here's the way I ended the day.

Narrator: The class applauds a picture taken at dusk from a hilltop, showing the whole town. Car headlights dot the roads, and lamps sparkle in the windows of houses.

Miss Greene: Brian, I think you created a very vivid picture of life here today.

Narrator: The class watches as Miss Greene solemnly adds the book to the box.

Charlie: I hope the class twenty years from now remembers to open our time capsule.

Brian: We can all come back and tell them about our soon-to-be legendary time capsule. Deal?

Class: Deal!

Think Critically

1. Why does Brian worry about choosing something for the time capsule?

2. How is a time capsule different from a museum?

3. How will people in the future find the time capsule?

4. What do you think the future fourth graders will think of Brian's photos?

5. Do you think time capsules are good ideas? Why or why not?

Art

Your Time Capsule Make a list of items you would include in your own time capsule. Then draw a picture of each thing. Arrange the pictures artistically on a sheet of paper. Include a border around the edge and a title.

School-Home Connection Talk to family members about what their lives were like when they were in fourth grade. Discuss how their experiences were like and different from yours.

Word Count: 1,249